Leaping Learners
Education, LLC

For more information and resources visit us at:
www.leapinglearnersed.com

Photo Credits

Fish diver graphics © Alexey Bannykh/stock.adobe.com
Cover © kondratuk/stock.adobe.com; Title Pg. © Andrea Izzotti/stock.adobe.com; Pg. 1-2© Richard Carey/stock.adobe.com; Pg. 3-4 © Pinosub/stock.adobe.com; © John Anderson/stock.adobe.com; Pg. 5-6 © anna_sea/stock.adobe.com; Pg. 7-8 © Alex Churilov/stock.adobe.com; Pg. 9-10© stigmatize/stock.adobe.com; Pg. 11 © Sevendeman/stock.adobe.com; © tetxu/stock.adobe.com; © Pakhnyushchyy/stock.adobe.com; Pg. 12 © cherylvb/stock.adobe.com; © AlexanderNovikov/stock.adobe.com; Pg. 13 © Vasileva/stock.adobe.com;

All design by Sean Bulger
All other photos done by Sean Bulger or royalty free from Pixabay.com

Special thanks to our reading level consultant
FRANCINE DEL VECCHIO, Ed.D.
Director of the Literacy Center
Professor of Education
Caldwell University

ISBN
978-1-948569-01-9

Dear Parents and Guardians,

Thank you for purchasing a *Fay Learns About* series book! After teaching students from kindergarten to second grade for more than seven years, I became frustrated by the lack of engaging books my students could read independently. To help my students engage with nonfiction topics, my wife and I decided to write nonfiction books for children. We hope to inspire young children to learn about the natural world.

Here at Leaping Learners, LLC, we have three main goals:

1. Spark young readers' curiosity about the natural world
2. Develop critical independent reading skills at an early age
3. Develop reading comprehension skills before and after reading

We hope your child enjoys learning with Fay. If you or your children are interested in a topic we have not written about yet, send us an email with your topic, and maybe your book will be next!

Thank you,

Sean Bulger, Ed.M

www.leapinglearnersed.com

Reading Suggestions:

Before reading this book, encourage your children to do a "picture walk," where they skim through the book and look at the pictures to help them think about what they already know about the topic. Thinking about what they already know helps children get excited about learning more facts and begin reading with some confidence.

Preview any new vocabulary words with your child. Key vocabulary words are found on the last few pages of the book. Have your children use each new phrase in their own words to see if they understand the definition.

After previewing the book, encourage your child to read the book independently more than once. After they have read it, ask them specific questions related to the information in the book. Encourage them to go back and reread the relevent section in the book to retrieve the answer in case they forgot the facts.

Finally, see if your child can complete the reading comprehension exercises at the end of the book to strengthen their understanding of the topic!

This book is best for ages 7-8
but. . .
Please be mindful that reading levels are a guide and vary depending on a child's skills and needs.

Fay Learns About . . . Octopuses

Written by Sean and Anicia Bulger

Table of Contents

Hi! My name is Fay. I love to discover and learn new things. In this book, we will learn about octopuses. Let's go!

Introduction

The octopus is an amazing ocean animal with special features you would not believe!

Body

Octopuses have a head and eight arms. An octopuses body does not have any bones. This allows them to fit into small spaces to keep safe. They have **gills** that let them breathe underwater, similar to other ocean animals.

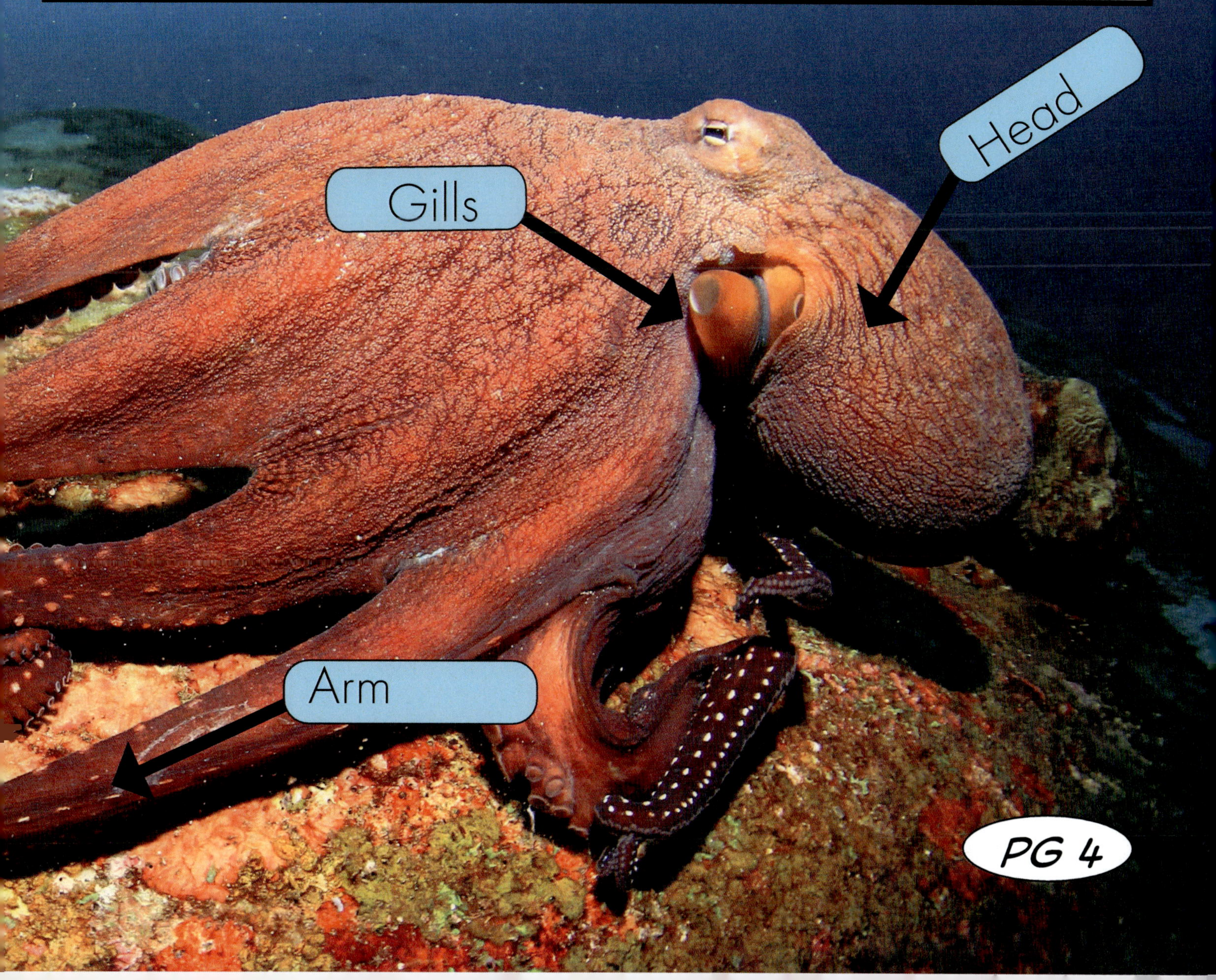

Octopuses have **suction cups** under each of their arms to help them move and eat.

Octopuses have a special ability to regrow their arm if it falls off or is eaten. This is called **regeneration**.

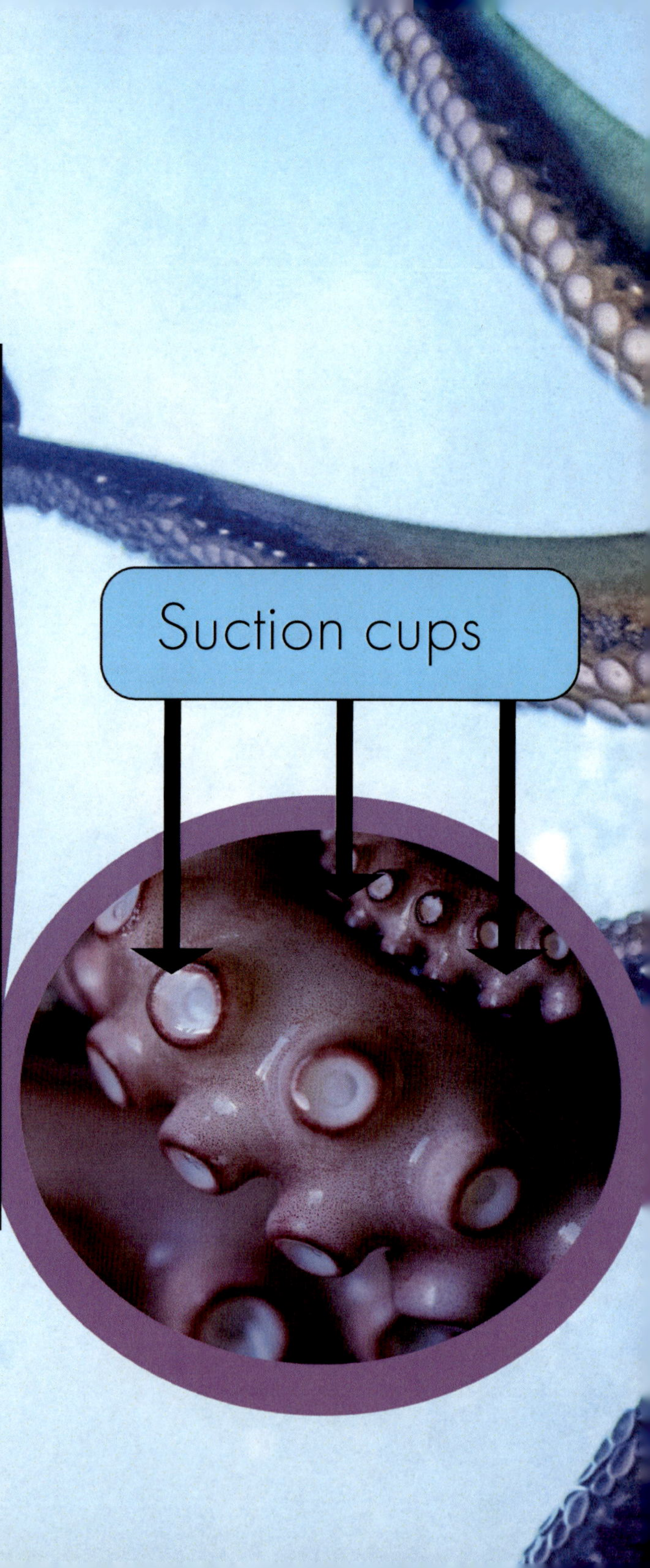

Tricky word:
Chunk it out
re-gen-er-a-tion

Protection

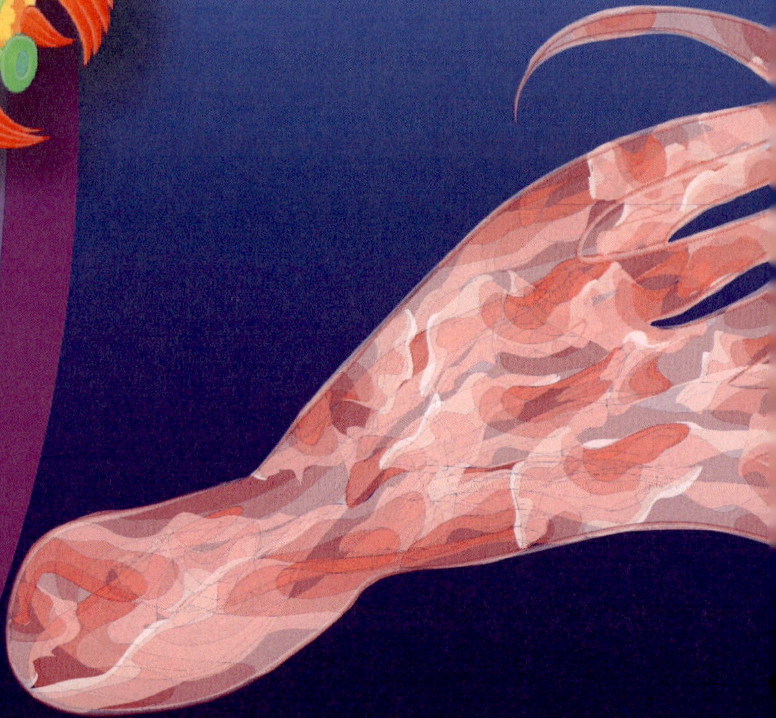

To protect themselves, octopuses will bite predators with their strong beak or shoot a cloud of black ink that swarms around the other animal. Then octopuses quickly swim away from danger. This makes it very difficult for predators to hunt and kill octopuses.

Can you spot the octopus?

Octopuses can blend into their surroundings by changing their skin to match the color of their habitat.

This **camouflage** helps them stay safe and catch prey.

Some octopuses are poisonous.

The blue-ringed octopus has one of the most poisonous bites in the world!

The blue-ringed octopus has enough venom to kill 26 adult humans!

Habitat

Octopuses can be found in all the world's oceans, but are commonly found in coral reefs or tide pools. They make **dens** to stay safe from predators.

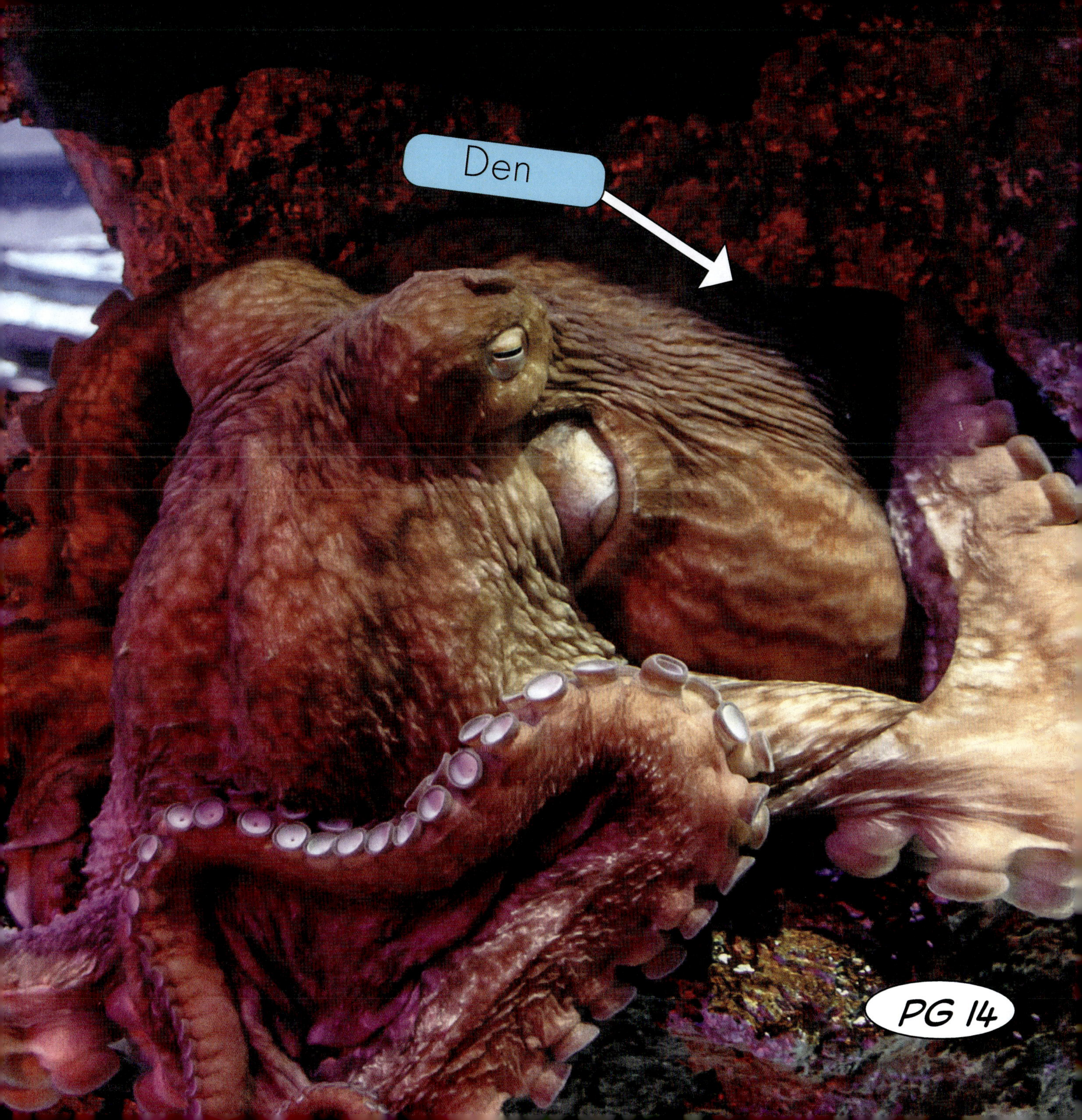
Den

Diet

An octopus's diet includes crabs, clams, and sea snails. It uses its arms to open its prey and eat it with its beak-like mouth. Some octopuses are poisonous and **stun** their prey before they eat it.

Snail

Clam

Crab

Predators

Octopuses are difficult to hunt because of their camouflage and speed. When there is a shortage of food though, animals such as sharks, dolphins, and eels eat octopus.

Eel

Dolphin

Shark

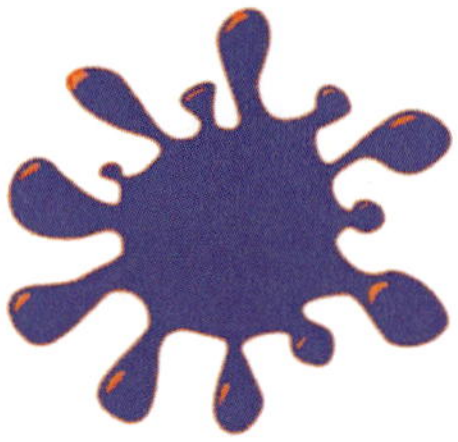

Octopuses have blue blood.

The heaviest octopus ever recorded was 600 pounds.

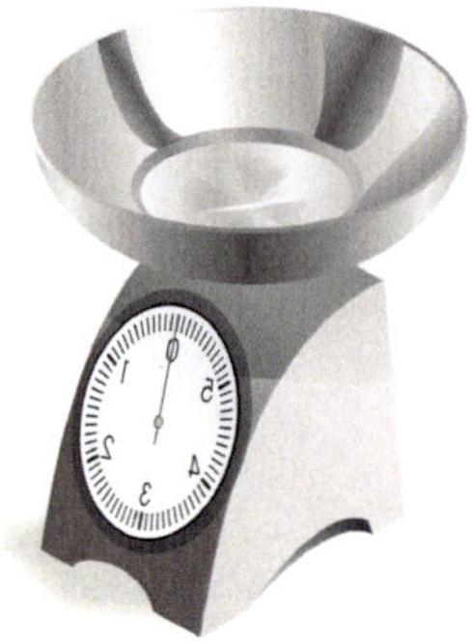

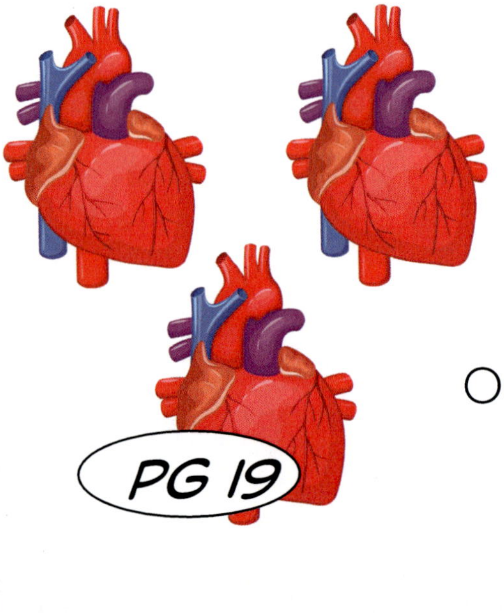

Octopuses have 3 hearts.

Glossary

A glossary tells the reader the meaning of important new words.

Suction cups – Cup like part of an animal, usually on their arms or tentacles. Suction cups grip things using pressure

Regeneration – To grow back a body part

Camouflage – To blend into the surrounding environment

Den – A hole or cave where an animal lives to keep safe

Stun – To be paralyzed for a limited time

Ink – A dark black liquid

Draw a picture of an octopus

Choose 3 words from the glossary and write a sentence for each one.

1.____

____

____

2.____

____

____

3.____

____

____

Quiz

1. How do octopuses fit into small places?
a. They are small
b. They move rocks
c. They do not have bones

2. Which animal do octopuses eat?
a. Snail
b. Shark
c. Eel

3. What is one way that octopuses protect themselves?
a. Jumps out of the water
b. Sprays ink
c. Runs fast

Common core standards:
RI. 2. 1 - Questions 1, 2,3

4. What is it called when octopuses regrow an arm?

a. Regeneration

b. Suction cups

c. Den

5. In which section do you learn about where octopuses live?

a. Body

b. Habitat

c. Diet

6. How many hearts do octopuses have?

a. 1

b. 2

c. 3

Common core standards:

RI. 2. 5 - Question 4

RI. 2. 6 - Questions 5

RI. 2. 1 - Question 6

Answer the following questions using information from the text:

1. What do octopuses do if attacked?

2. Where do octopuses live?

3. How does camouflage help octopuses?

4. Which animals eat octopuses?

Want to learn about rainforest animals? Check out the "Matt Learns About..." series!

Great for emerging readers ages 6-8

Want to learn about colors? Check out the "Clayton Teaches you About..." Series!

Great for early readers ages 4-6

Want to learn about Farm Animals? Check out the "Katie Teaches You About..." Series!

Great for early readers ages 4-6

Made in United States
Orlando, FL
02 November 2026

90104234R00024

Made in United States
Orlando, FL
02 November 2026

90104234R00024